Birds at the Beach and Beyond

Lisa Donovan

illustrations by Dee Paras

ISBN: 978-0-578-27622-9

Printed in the United States of America
Signature Book Printing, www.sbpbooks.com

Layout by Van-garde Imagery, Inc.

Dedication

With thanks to the Creator for birds
and all that they teach us.

Preface

Birds live all over the world — just like people do. Some birds live near water. Not just ducks and geese, but shorebirds. Seabirds. Wading birds in a wetland. You see them near a huge ocean, a muddy marsh, or even a small pond at a park. Many birds fly for days to be in warm, wet places. Other birds live by water all the time. Let's see...

Birds near the water
Are wonders — just look!
On sand or in mud,
By marshes or brook.

You can't get too close.
Just watch from afar.
Soon you will see how
Marvelous they are.

Pelican parades
Pass perfectly by.
They fly in a line
Along the sea sky.

Red Knots — world travelers
Who must rest and feed.
They probe sandy shores
To find food they need.

Ruddy Turnstones — yes!
That's just what they do.
Flip small rocks over.
Find one bug or two.

Sanderlings play games
With waves — so well planned.
Run from the water!
Find food in the sand!

Oystercatcher's beak
Looks like a carrot
Stuck on a bird head.
How can he bear it?

A **Black Skimmer**'s bill
Feels fish in the sea.
He finds one—then SNAP!
One less fish for me!

When you unpack snacks
To munch at the beach,
Keep them away from
The **Laughing Gull**'s reach.

The Royal Tern's crown —
A mess! Did he rush?
Get ready too fast?
He needs a good brush!

Piping Plovers dance
When trying to eat.
While hunting for worms,
They wiggle their feet!

Why won't the Willet
Show off her disguise?
Plain on the ground, but
Bold stripes when she flies!

Least Tern has the most
Secret kind of nest.
A scratch in the sand
Hides her eggs the best.

Surprise! Sea Turtles!
Not birds, but it's grand
They too can lay eggs
In nests under sand.

Swoosh goes the Osprey
Feet first for a fish,
Then home so the chicks
Can eat it — delish!

Does a Black-Necked Stilt
Ever stop to think
That her long, thin legs
Are bubblegum pink?

If you were tiny
And had lots of nerve,
You could slide right down
White Ibis beak's curve.

Tricolored Heron —
An amazing sight:
Blue, gray, and purple
And bely bright white.

Dinosaur head with
Long beak, legs, and wings,
Wood Storks are surely
Quite curious things.

The **Great Egret**'s neck —
A big letter "S."
It turns into "I"
When stretching I guess.

Anhinga's long neck
Is thin as a stem.
When drying his wings,
He makes a big "M"!

Great Blue Heron waits
All hunched in the rain,
Then stands tall and still
When sun shines again.

See Spoonbills zigzag.
In water they work.
They look for small fish
Without knife and fork.

If Snowy Egret
And you were to meet,
I bet you'd stare at
His bright yellow feet!

The light slips away
From land and the sea.
Then dark quiets birds
And sets the moon free.

Tomorrow will start
When night's turn is done.
Sleep now and dream well.
Birds rise with the sun.

Be a Friend to Birds at the Beach and Beyond!

Always keep your dog on a leash.

Look for signs or fences that tell you where birds are eating, resting, or nesting. Please don't touch nests or go in these important areas.

Give birds a lot of space — never get too close. Use binoculars or spotting scopes to see birds.

Don't scare birds and make them fly away.

Always throw away your trash, including fishing lines.

Keep drones, kites, and fireworks far away from birds.

Want to learn more?

Visit the National Aviary

Located on Pittsburgh's historic Northside since its founding in 1952, the one and only National Aviary is home to 500 birds representing more than 150 diverse species from around the world, many of them threatened or endangered in the wild. The National Aviary's large walk-through habitats create an intimate, up-close interaction between visitors and free-flying birds, including opportunities to hand-feed and to meet many species rarely found in zoos.

NATIONAL AVIARY

PITTSBURGH, PA

National Aviary • 700 Arch Street • Pittsburgh, PA 15212 • Phone: 412-323-7235

Many thanks to...

Katie B. Barnes

Coastal Stewardship Manager, Louisiana
National Audubon Society

Executive Director Jim Bonner and Staff

Audubon Society of Western Pennsylvania
Pittsburgh, PA

Melissa K. Chaplin

Endangered Species Biologist / Recovery Team Lead
U.S. Fish & Wildlife Service
South Carolina Field Office
Charleston, SC

Aaron Given

Wildlife Biologist
Town of Kiawah Island, SC

Dr. Millie Myers

Professor Emerita, Tepper School of Business
Carnegie Mellon University
Pittsburgh, PA

Adriene Rister

Owner, Spark Books
Aspinwall, PA

Janet Thibault

Wildlife Biologist
South Carolina Department of Natural Resources
Charleston, SC

With gratitude for expertise, research, suggestions, and encouragement to...

Robert Mulvihill

Ornithologist

National Aviary

Pittsburgh, PA

Bette Popillo

Kiawah Island

Shorebird Stewardship Program Coordinator

Kiawah Island, SC

Much love to...

Bill, Annie, and Claire Donovan

Experts at yelling, "Look, a bird!"

About The Author

Lisa Donovan is very good at looking out the window at birds, and that's why she wrote **Birds in My Backyard** in 2019. This book, her second, proves that she also loves looking at birds at the beach, in marshes, on ponds, and by lagoons. She encourages others to join her.

Lisa grew up in Greensburg, Pennsylvania, and graduated from Grove City College. She received an M.A. in Professional Writing from Carnegie Mellon University. After working in communications for various companies and clients, she volunteers and looks at birds.

Follow Lisa at www.ldonovanbooks.com.

About The Illustrator

Dee Paras remembers kindergarten teachers trying to lure her away from the crayon table. All she wanted to do was her art — and she loved color! Later in life, Dee worked with sight-impaired children and learned to read and write in Grade 2 Braille. Her students inspired her to create simple images that were visually appealing. She lives as a creative, reveling in various media, from watercolor painting to collage to knitting.

Dee grew up in Green Tree, outside of Pittsburgh, Pennsylvania. She graduated from Grove City College with a B.A. in Business Administration. She is pursuing her dream of being a children's book illustrator and working on art projects full time. This is the second book that Dee has illustrated.

Follow Dee on Instagram and Facebook.

The Illustrations in This Book

Dee Paras created the illustrations using collage. Collage is art that is shaped by arranging pieces of paper, photographs, and other materials on a support (wood, canvas, paper, etc.) and gluing them down.

Dee first sketched each illustration and added value. Value shows which parts of the page are lighter and which are darker. At this point, she also consulted with an ornithologist to make sure the drawings, while not scientific representations, included generally accepted characteristics of each bird.

She then painted papers of different textures and weights with colorful acrylic paint. When the paper was dry, she cut it into shapes and arranged the pieces to form the pictures that tell the story.

For examples of different papers with different textures, go back in the book and look at:

- The sky on pages 4 and 5
- The Red Knots' wings on page 7
- The beach on page 9
- The ocean on page 11
- The eggs on page 16
- The nest on page 18
- The water on page 19
- The grass on page 21